ALL THE EMERGENCY-TYPE STRUCTURES

ELIZABETH CANTWELL

Inlandia Institute

ALL THE EMERGENCY-TYPE STRUCTURES

ELIZABETH CANTWELL

Cover art by Kenji C. Liu
Book design and layout by Kenji C. Liu

Printed and bound in the United States
Distributed by Ingram

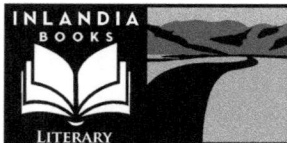

Published by Inlandia Institute
Riverside, California
www.InlandiaInstitute.org
First Edition

ALSO BY ELIZABETH CANTWELL

Nights I Let the Tiger Get You
Premonitions

for C,
my storm and roof

For the protection of the sun a shelter must be built: first for the bottle, then for bread, a sausage, and for a man; then for two or more people, in which later they can cook and even spend the night.

—B. Juvanec, "Stone Constructions in Corbelling"

TABLE OF CONTENTS

I. EARTHEN

SITUATIONS OF CASUAL DANGER 1

EVENTUALLY AN EXPERT CAME, HOODED HER, AND TOOK HER AWAY 2

IT IS DIFFICULT TO ABANDON OBLIGATIONS BUT IF YOU'RE
GOING TO DO SO YOU'D BETTER END UP IN THE CALIFORNIA DESERT 3

IT'S WEIRD AND PISSED OFF, WHATEVER IT IS 4

KEEP IN MIND, ZOMBIE BEES ARE VERY RARE 5

HOUSEWARMING 7

PREPARATIONS 8

HAVE YOU EVER WONDERED HOW MANY PEOPLE HAD THE SAME EXCITING
DREAM LAST NIGHT AND HOW LITTLE, THEREFORE, IT MEANT 11

II. AQUATIC

THE PEOPLE WHO LIVE IN BOATS 17

IT WORKS IN ANY SPACE, FROM BEDROOMS TO KITCHENS 18

DID YOU HEAR THAT, THIS GUY THINKS IT'S BEEN TERRIBLE 20

WE ARE ALL CAPABLE OF ENLIGHTENMENT, OF HASTENING OUR OWN DEMISES 21

ISOLATES 23

HOW COME NO ONE ON TWITTER IS TALKING ABOUT THIS 24

SPILLWAY 25

III. ASTRAL

SOMETHING NOT EMPTY 31

WEAK STARS 33

IT'S THIS AGAIN 34

YEAR OF THE BEES 35

WE WILL GO HOWEVER FAR IS NECESSARY TO MAKE IT COHERE 37

METEOR SHOWERS 39

IT'S PRETTY FUNNY WHEN YOU THINK ABOUT IT 41

YEAR OF OUMUAMUA 43

THE MYTHS WE INHABIT 44

IV. ENCLOSED

THE TRUTH IS OUT THERE 49

ANOTHER INEXPENSIVE SOLUTION WITH A BIG PAYOFF 50

I WOULD LEAVE MYSELF INSIDE A GLASS CASE IF POSSIBLE 52

YOU SHOULDN'T WRITE A SONNET WITH A MACBOOK IN IT 53

THIS IS AS ACCURATE AS I CAN MAKE IT 54

DAILY COMMUTATIVE PROPERTY 55

ATOMIC LICENSE 58

READ THIS WHOLE THREAD 60

I DON'T HAVE AN IPAD 61

EMERGENCY QUEEN 62

V. ABLAZE

MATINEE 68

A RICH HISTORY OF MEN PAINTING THINGS BLUE 69

THAT WHICH IS FORM IS EMPTINESS, THAT WHICH IS EMPTINESS FORM 70

THESE WEREN'T A FEW BIRDS 72

FROM A DISTANCE IT COULD BE A SLOWLY BLOOMING FLOWER 73

THE RESEARCH WAS RECENTLY PUBLISHED 74

IN THE POST-WAR ERA, IMAGES OF WHEAT FIELDS WERE COMFORTING 76

NOTES 77

ACKNOWLEDGMENTS 79

I. EARTHEN

Landscape does not exist without an observer, without a human presence. The land exists, but the 'scape' is a projection of human consciousness, an image received.

—Gretel Ehrlich, "Surrender to the Landscape"

Shelter carries the greatest immediacy in developing plans to overcome a disaster situation. When considered in terms of the rule, 'shelter' doesn't necessarily refer to a structure, or as it's often put, a roof over our heads.

—James D. Nowka, *Prepper's Guide to Surviving Natural Disasters*

SITUATIONS OF CASUAL DANGER

On the question of skinny jeans, science has mostly made clear
they will not kill you. Blood clots seem not to be
a threat. Set down by the side of the bed, dormant, they even appear

welcoming, safe. So few carefree places
remain: the smart phones could catch fire,
the bookshelves could peel from the walls, the snap peas

could harbor e. coli, the car tires
could begin rolling without warning. In such situations
of casual danger, one must be on guard against desire.

Do not touch the flesh like that. Do not allow a dilation
of the pupils. Put down the body and collect
a pole, some tree branches, pine needles, enough vegetation

to heap two to three feet deep across the protective,
rib-like sides of the structure that will house you.
If you find yourself short on brush, you may erect

a shelter using only carefully chosen language. The view
outside may be frightening: giant beetles devouring
water drops, a man in a wolf's mask carving honeydew.

Don't panic; this will soon seem like home. The wildflowers
are so lovely here. And the words, peeling off your skin
in ribbons, weave their mesh nets stronger by the hour.

EVENTUALLY AN EXPERT CAME, HOODED HER, AND TOOK HER AWAY

I want to tell you about the new planets.
How one of them experiences a whole year
in just a day and a half. Ray Bradbury shit,
you know, gravitationally locked, with one side
always in the dark. The other side bathed in
salmon-colored light, a permanent sunset
feeling, the quiet star, an alien waterfall.
I want to tell you about the desert flower
phenomenon, the year of more rain than the
barrels could hold, the sand verbena and white
primroses now poised to carpet this granulated
and angry land, to smother it in purple and green
and white.
    Impossible landscape:
nowhere to step but it blooms. And the hawk
that landed at some point in the late morning
in the middle of the campus fountain, standing
claw-deep in the green water, the goldfish
swimming around her feet, darting white and
orange underneath, as she almost imperceptibly
blinks her eyes. Perhaps injured, or digesting
some rodent who'd been poisoned; perhaps
just hot, and wary of the ring of people around
her, holding their small electronic devices up
to her eye level and tapping. Hawks of her kind
have yellow irises until three or four years
of age. Here I am, I want to say, circling slowly
around a budding sun, holding all this alien life
in my belly: exotic, undigestible, ready to burst.

IT IS DIFFICULT TO ABANDON OBLIGATIONS BUT IF YOU'RE GOING TO DO SO YOU'D BETTER END UP IN THE CALIFORNIA DESERT

You called but I didn't pick up.
Dissonance: the body a channel for sound.
Down down in the deep heart of you.
A liquid sting, a solitary boat spidering out at noon.
The bodies asleep in the wheeled chamber, wheeling through space, space as the
    medium, space as dissolver and hand.
Mountain, mountain, sand.
The mouth to the faucet, the faucet made of sand, the hot whatever bubbling up into
    the throat, smoke, saxophone.
If the world outside were burned into the land would it hurt.
Humanoid trees twisted, unplanned.
The brain is ajar on either end and what flows through it makes it different.
Far off in the wind, something opening up.
A box made of sand with sand inside it.
The top of a bird's head.
A granary, vacant for many years now, filled last night to the brim with polished
    stones—all colors, all sizes, all heavier than a grown man's heart.
I would call you but the type of call I could make is audible only to dogs, seaweed, a
    certain breed of brain cell.

3

IT'S WEIRD AND PISSED OFF, WHATEVER IT IS

The dry bleached mountainsides waiting to ignite.
Don't fight it, it's supposed to burn.
We've been waiting here for our drinks for entire minutes.
If you are feeling overwhelmed tonight by
contemporary life
there is a QR code in the kitchen you can scan.
Really, it's our own human presences that have forced
the natural rhythm of things to this unlit staccato.
Don't ask me what it does; I haven't tried it,
I lied.
I ate the piece of rotting cactus someone threw
over the fence.
I don't remember the last time he kissed me.
What we wore when we decided it was over.
The mountains again: fuck you
for wanting to conquer them, their symbolic whatever.
Did you hear those clouds.
I wanted to write a poem about Kurt Russell
in *The Thing* but this is the best I can do: cold and far
away, perhaps inhabited by something other,
staring across the landscape and awaiting
the quiet end no one feels good about.

4

KEEP IN MIND, ZOMBIE BEES ARE VERY RARE

Zombie Bees: Can They Infect Us?   This is a real headline
I really read just now   There are these bees
that get infected by parasites and start flying aimlessly

at night   Drawn to bright lights   Dying
within days   the parasite's larvae emerging from the dead
body under the moon

like some beautifully fucked-up B movie shit   If you find
a body you believe
to be infected you are supposed to enclose it carefully

inside a Ziploc bag   You are supposed to dispose of it
with caution   Outside now   standing in the
village   waiting for people to swarm through

staring at their tiny screens and anticipating small
brightly colored   animated beings
I think about the fights we have over and over again   The

whiteness seeping through the flags
hanging in front of houses   On all the pages
of all the newspapers   red hats   closed eyes   Peculiar power:

the inability to see those centuries
of blood and ropes and twisted fields we told ourselves we'd
long ago collapsed into a book about the dead

now rising back up around us whiter and sharper and
uglier than ever   In front of me   a bee I took for dead
stumbles up   flies straight into

a resumed life of queens and honey   I put
the sweaty Ziploc bag in my hand away   balling it up
like a used Kleenex   wondering

what level of fear a bee is capable of

experiencing   To smile through this
life   is to accept unquestioning a certain number of

repulsive burned-out set pieces   I think I have to inhabit
something angrier now   I think
I need an uneasy phrase to yell out   A stretch

of displeased space   I have pieced it together   this
communicable dread   our sole emotion   If you want me
you will find me dancing robotically in the streets

HOUSEWARMING

The garden the previous tenants left is abandoned
save for a handful of peppers thumbing tenacious flesh
at the heat   I roll our trash cans next to them   I am replete

with June   High noon   I am convinced there is something
medically wrong with me   My voice has been hoarse for a month
My digestion off   I wake up in the early hours of the morning

with my heart beating like a hundred clocks   The sun is a heavy hot
smock curling over my shoulders   There is a pocket inside each
of us that   pushed sharply enough   could pop   could

ask us to take the knife out of the kitchen and plunge it into the dirt
until it hits bone   To feel that sick sad
yes   Severed roots hissing through the mess

in the palms   Let us now undress the world   Let us peel off
its crust   its mantle   its outer core   Let us find the poor sore
soul at its center   Covered over in grief and triggered and worn

to its own tiny world bone   Let us reach in and draw it out
through the blood and muscle and pulsing skin   Transplant it
somewhere inside ourselves   Holding on   Lying down

in the middle of it all   Tall tall bodies exposed
to space   Only then will this place feel like home

PREPARATIONS

The humpback whales off the coast of South Africa
are amassing in numbers the scientists
can't explain   Usually humpback whales

are a solitary species   Two or three or four
might swim together for a while
then go their separate ways   little polished

stones drifting off to be alone   The stones my son
collects from other people's landscaped lawns
on his walks around the neighborhood

bump together in the plastic bag I know now
to bring with us   Two weeks ago   he and my
husband came home

telling me about a dead tree just
down the street   They'd picked brown twigs off it like
hangnails   skin with no feeling   seeing nothing

green inside   Recently the whales
have been spotted in "super-groups" as large as
200   gathering in tightly knit pods the size of

football fields   Perhaps there has been a change
in the availability of prey   Perhaps this
behavior has always been innate

but has at last been made possible by a
resurgence in the humpback whale population   I ought
to be encouraged by this

oceanic proliferation   Instead I have
the urge to build a bomb shelter   What do the whales
know that we don't   Who can I turn to

to form my tribe
if it turns out I need one after all   Why
are so many emergency-type structures

built underground   What if an emergency makes being
underground a little less
than helpful   Some movement from the

belly of it all   Today I walked our dog
past the tree on our block and it was full of
green leaves   I took a picture of it   Have you heard

about the forest in Poland
where all the tree trunks bulge out markedly
toward the north   Perhaps there was a heavy snowfall

one year while the trees were still young   weighting the
incipient trunks with ice   hard-packed heavy white blocks
darkening out the sun   Perhaps some farmers

in collusion with boat-builders or furniture-makers
spent their time while the trunks were still malleable
carefully crafting the perfect natural curves

for the bow of a boat or the back of a dining room
chair   and the curves themselves became some sort of warped
genetic inheritance   passed down in the DNA of the soil

itself   What was dead is not dead   What was waning has
proliferated   The skies have begun to exhibit
a new kind of particularly unsettling cloud formation

A new son is somehow growing lungs inside my body   I think
I can make him part of my tribe if I have to   Perhaps
there has never been a natural collection

more integrated than the hodgepodge of sticks
and fallen blossoms and insect carcasses and stones
knocking together in the bottom of my son's

dirty plastic bag   Are you ready for the impending
blossoming that will have no causal explanation
and will likely blanket us all   Spiraling inward until even the

most remote underground shelter   packed to the brim
with gallons of water   canned peaches   flashlights   linens
finds itself full to rippling with unexpected life

HAVE YOU EVER WONDERED HOW MANY PEOPLE HAD THE SAME EXCITING DREAM LAST NIGHT AND HOW LITTLE, THEREFORE, IT MEANT

We've been afraid to acknowledge the lack of seasons for some time now.
The mornings unable to get out of bed, hit with the coldest stream
of water. The cells in the body's immune system taken out, toughened, taught
to stop at nothing, injected back in, sent off into the dark streams to fight it
out. The scream of outer space, its face, how close it is to our patios. Planes
crash, mothers die, the infant is brought into the house and breathes it
down with his fire. And there is the muddled mundanity of the
daily routine—the showers, the chopping boards, the highway slowdowns, the
unexpected surges of desire. This is a face I cannot stop touching,
even on the edges of the most upsetting cliffs. Do you see it
now? The world rotating itself into morning—It's like an ecstasy
no one wanted to admit. The earth rent open; us, kneeling, molding it
back together with our mouths. The well of light our hands turn into
as we walk across small hills, carrying glass bowls, refracting the sun.

II. AQUATIC

When you experience beauty, you experience evidence in your inner
space that at least one thing that isn't you exists ...
The basic issue with beauty is that it is ungraspable. I can't
point directly to it and I can't decide whether it's me or the
thing that is emanating beauty. There is an undecidability (not
total indeterminacy) between two entities—me and not-me, the
thing. There is a profound ambiguity. Beauty is sad because it is
ungraspable; there is an elegiac quality to it. When grasped, it
withdraws, like putting my hand into water. Yet it appears.

—Timothy Morton, *Dark Ecology*

You should have stayed underwater.

—Nord, in *Waterworld* (1995)

THE PEOPLE WHO LIVE IN BOATS

It is easy to become one of the people who live in boats. You can actually make the decision yourself—all you need is a boat and a body of water, unless you want to be one of the people who live in boats in their driveways, which is certainly the less attractive option. There are many rules having to do with living in boats. The people who live in boats must not know how to swim; otherwise, there is a danger that one evening, looking down deeply into the black mirror shimmering against the side of their permanent residence, they might realize it looks like the dream they have been trying to get inside of for years, they might imagine it the slowly opening bloom they have held their breath waiting to smell. The people who live in boats must have to winnow their possessions. The people who live in boats perhaps did not have that many possessions to begin with—knowing that they were the sort of people who would end up living on boats, they steered clear of the bookcases, the porcelain figurines, the baseballs in their clear glass cases. The people who live in boats must love whiskey, hate taxidermy, feel largely indifferent towards types of lighting fixtures. The people who live in boats must have ears that look blue in the moonlight, and if you embrace one of them, your shirt must come away feeling damp. The people who live in boats are all around us. If you were to live in a boat, you would wake up one morning and realize that the skin on your face was slowly turning a slick, cool gray. No one would be able to tell from behind. But in profile, it would be quite obvious: the translucency, the minnowing, the eye with the inner lid. There is a river in New Zealand that was just granted the same legal rights as a human being; the tribe that fought for its human recognition believes strongly that it is one of their ancestors. *We can trace our genealogy to the origins of the universe,* says the lead negotiator for the tribe. *And therefore rather than us being masters of the natural world, we are part of it.* The people who live in boats must become familiar with the blurry boundary between kin and predator. The people who live in boats must travel far but stay in the same sphere. The people who live in boats must curl up every night in their gently rocking beds and, growing silently bigger in sleep, dream the globe alive.

IT WORKS IN ANY SPACE,
FROM BEDROOMS TO KITCHENS

I have spent some time wondering
about the condensation gathering beneath
my legs at night   I have draped my window
with ocean print cloth   It's a very versatile
print   consisting largely of red octopuses and
chartreuse kelp and a single off-center
jellyfish   Sometimes I look at the jellyfish
It seems off to me   I touch it
It feels like nothing   I worry that
my fingers are too rough or angry   I think about the
jellyfish   injured   assessing the
damage   finding itself lopsided   lining up
the crosshairs of its floating mind with
some intended central
point

It once had a home   It once took note of a
butterfly   The jellyfish has had no water
for a long time   It feels cold   It feels
very far away   It is old   older than you know
How cold do you think you would have to be
before your fingers forgot how to
work   Should you try it   Should you really
get that deep   Should you have
another latte   another selfie   Should you write
that blog post   How much time
before you have to meet the guy at the hip
experimental art bar   Do you think you could
get cold enough before then   Do you think your skin
could go all blue and transparent   Do
you remember how to stop the
noise

The infinite imagined to be curving in on us
all the time   There was a ghost horse sent up

into space   Galloping in an
arbitrarily large radius   You heard it once
below you   As if projected onto an underground
inverted screen   *We observe the sky as it appears*
blue and cold and without
sin

More of me and more of me and more of
me and more of me is seeping into the land   My hand
raised up with nothing in it   It comes from the ocean
There is too much of it to weigh or count
Sometimes it can build up inside a person like a
storm   It can make your flesh hard
It can be used to mark the difference
between silence and awake   If you would like
to stay here I have built you a
display case   You cannot grow at all here or learn
the alphabet or the stages of grief
But you can love me   And the tide will never
turn

DID YOU HEAR THAT, THIS GUY THINKS IT'S BEEN TERRIBLE

There's something about the man in the bathrobe wandering around
in the dark   A purple sky in the windows   A lark that tried to open
too early   He can't find the light switches   he's tired   the TV is on too
loud   Walking straight forward like somebody possessed   like somebody
walking into the ocean on purpose   The salt collecting around the edges of the
skin   A thin spine trying to hold flesh together like a giant metal claw in
a cheap arcade   The form so isolated in space   A crying emperor in the
middle of an empty field   But oh   how surrounded we are with bodies   A tide
of bodies   We can barely move forward because of the bodies   our son
clings to your neck   I'm sweating   I'm scared we might be
happy   Everybody's waiting for the reveal   Out there   over heads   a
wave swells on the horizon   one we know we're going to have to duck
beneath   one we hope we have enough breath in our lungs to brave

WE ARE ALL CAPABLE OF ENLIGHTENMENT, OF HASTENING OUR OWN DEMISES

You were wrong about the mammoths. They didn't give up
on some snowy plain, tired and old, hunted by
The Other. You should know this, the way you pride yourself

on the truth. If you imagined them scared, sad, dying alone, arrows
in their sides as a man stood over them: you were wrong.
Those mammoths in their taxidermied displays. Their

tusks. You should know: a group of them
kept living on a tiny island long after their fellow mammoths
died, millennia passing, the last few herds swelling over the

hills. The crater lake like a finger pushing slowly
into the flesh on the side of your
neck. A still drop of silvered mercury

suspended in the air. The sea seeping beneath the isle. You were
wrong when you imagined me scared, sad, not yet sure
how to navigate the topography. How little

you knew. The honey dripping off the tips of my fingers
like blood. In the picture I am not even
smiling. The ocean stretches out behind me like a

jewel trapped in a case. I'd spent the whole night
with my eyes wide open. There is something inside
that wants to fall endlessly through swaths of purple velvet,

to land on its back hard, to black out, to come to
in a mess of limbs, to forget the names of the
children. It's dark and I know exactly how to be here.

The lone shallow freshwater lakes shimmering and shrinking
in the hotter north air. The water saltier and saltier. The mouths

more and more desirous. You must have guessed

by now. We are all agents of our own
destruction. You, camera in hand, unable to see me as anything
but a child. The giant feet, anxious, tramping down the banks

by the last cool pools, tracking, a ticking clock, knocking away
the minutes left. The brackish water rising. The light in the eye, the
paradise of feeling your brain expand until it touches

the other side. You should know by now that your rush to suck
the last sweet drops from the surface
could only have made it worse. The touch on the arm. The

dried salt crystals left behind. The picture folded at the bottom of the
secret suitcase: an island heavy with dead bodies, aching for diamond
streams, for heavy breathing, for the chance to do it again.

ISOLATES

There must be at least one square foot of the surface of the ocean
that has never come into contact with anything   Not a dolphin
surfacing for air   Not a ship sailing to the edge of the known world before
there were maps   Not a pelican skimming for a midday snack   No man
thrown overboard thrashing around   No ark   No ash   No minor prehistoric
shark   One perfectly dark and unbreached stretch of sea
There must be at least one tree whose leaves have never seen
a sunrise   Whose bark has no idea what lichen feels like   Whose hiked-up
roots shoot through no naturally trodden paths   This is where you take me
these untouched places of the world   swirling eddies   perfectly lone
inside themselves   Our own square foot of dearth   the purest swath of earth
cut off from any knowledge of a context larger than this jagged edge of paper
we are ripping   slowly   with our sharpest heart-shaped smiles   Here   I've found
yet another way to wall myself off   as you build so silently   stone by stone

HOW COME NO ONE ON TWITTER IS TALKING ABOUT THIS

The eye a camera humming across the top of the earth : cavern
gully   ravine   gorge   Far below   a tiny man jumps into a tiny pool   Black
pool   wet pool   pool pooling around itself   a
shimmering thing in the night   Did I mention the night   Hot and dry
on our faces   Another word for *dry* is *incommunicado*   I am not
going to apologize for this
anger   This eye like a knife cutting through the hair : gray   gray   white
underneath the brown   Bone sticking out oddly   A small mesh screen between
ourselves and the truth   the truth holding onto an orange phone   repeating
some version of *I didn't do it*   I know *we are running out of time* is a cliché
so I try to re-phrase it   to say it some way you'll understand : *have you seen*
our dog   the way he worries his paws   the dry
skin like a snake on his belly   I know you care about the dog   But when I go outside
to find you   to ask you if you'll bear witness to our sky's empty   convulsive
grief   you're nowhere to be found   The june bugs hitting themselves
against the siding   The eye still performing its fruitless
scanning   The somethings
whistling roughly through the tops of the palms   And overhead   a dark pool
gathering   dropping nothing of itself   refusing to yield even one syringe full
to the yellow door growing bigger and more luminous in our minds

SPILLWAY

My father never warned me against the slow, inward retreat.
The water spiraling into the bellmouth.
The woman who, passing by, finds herself fixed to the spot.
Who ever heard of such undisrupted motion?
And all of it pulled down somewhere we can't see.
Impossible blanket of fuschia around the base of a gray tree.

III. ASTRAL

You are an aperture through which the universe is looking at and exploring itself.

—Alan Watts

SOMETHING NOT EMPTY

I know people who get great anxiety imagining
what it would be like to float through space,

real space, untethered, inside of nothing, just
totally fucked. Like the Sandra Bullock movie,

the shot of the small white body and the
small white shocked face flying rather than

floating out into empty emptiness. When my
students gave a presentation to the class

describing how the Soviet Union wanted to
blow up the moon, I thought they were joking,

I even laughed. *Why would you want to shoot
nukes at the moon*, I said, and then, *Sorry, keep*

going. But it wasn't a joke. I read about it later,
one of those articles everyone passed around

online, a good read, an examination of what seems
to be a common paranoia about the moon,

that it could attack us somehow, that it's a
Nazi plot, that it doesn't give a shit about us,

that it has been trying with its moon face
and its moon pulses and its moon mouth

to make us crazy since the day we arrived here,
coming of course, from a myriad of other planets,

stars, galaxies, meteors, particles formed from dust
coming off of collisions—I never really understood

that, but Carl Sagan says it somewhere: we came

here travelers from the void, aliens, all of us, down

to the microbes cleaning our eyelashes. There
is no shape to space except our own slight

bodies. The ones we wield against each
other, as if we were even important enough

to be evil. As if the eye itself wasn't a sphere
in the night. *What is the grass, anyway*, I said

out loud to no one, ripping it right out of
the ground, shooting it into orbit, waiting

for its small green face to register the fear.
But all I could see was a light green relief,

a sense of moving quickly and at last towards
something not empty but chock full of blood.

WEAK STARS

There's this app you can get that shows you all the constellations
when you point the phone at the sky it identifies them for you   lit up on
the screen with little lines connecting all the tiny digital dots   but whenever
you're somewhere you can actually see the stars you likely won't have the
reception or the 3G or the LTE or whatever the app needs to function correctly
someone should make an app for conspiracy theories   when you point the phone
at a person it tells you what conspiracies they believe   JFK killed Stephen King
Elvis wrote Shakespeare's plays   contrails are chemicals designed to subdue
us   you could click on the little bolded theory name & follow links to learn
all about them   this is just a suggestion to anyone out there who could be
an app developer   in any case the sky is truly hung with bright things   with
things that could persuade you to take off your clothes for them in a bathroom stall somewhere
that could ring bells for you in the dark   the whole arc
of whatever spewing out after them   their charm quarks and strange quarks
dancing in front of you in all the blown-out Mexican restaurants of middle
America   the brightest ones (Sirius   Canopus   Alpha Centauri   Arcturus) look
from here like smug little prayers locked up in velvet boxes   lives blocked out
in charts   in earth   in apps   do you believe in the garden of Eden   in the possibility
that it might have been on Mars & Noah & all his descendants migrated here
across the empty vacuum   a flood of bearded floating spacemen   do you believe
in the injured jellyfish   rearranging symmetrically in their beds   do you
believe in the body   in golden numbers   in holding on to whatever it is you're spotting
as the head whips around & around & around in space   the woman in the chair in the
corner pounding her cane on the floor   yelling *pirouette pirouette pirouette*   & the
brain inside the whirling head believing the body to be just a trick of light   just a
garden something else ought to be growing in   tight orbits   tight
shoulders   the brightest stars in the sky are among the least studied

———

33

IT'S THIS AGAIN

I think I was in love with Mozart in another life   these recycled sunrises   these
tired trees layered limply on top of their own ghosts   out of the corner of my eye
a shoreline   and out of the shoreline   nothing   suppose you could
touch it   suppose you could try: that moment you knew there was something

to be tasted   that it was worth it   the ragged edges of forced laughs   the filtered
image of the couple   some nightmare version of us   on the beach   with those
stupid hats   suppose
you could make your mouth into the shape of me

in my mind I have devoured all the screens   I have seen us
together in unlit rooms   the perfume of something irresistible creating
a seal around us   mesh around tangerines   rotting tangerines   soft when you
touch them   too sweet   I have seen your lips on mine   and

in the mountains impassible tides   and in the tides
phosphorescence
the world I can imagine you in is all streetlights   weird pine trees   you would
never have climbed one   you would have waited for me   you would have

climbed one but only halfway   you would have spent a long time looking
for it (the right one)   you would have climbed it in order
to come down   you would have filled entire hours with someone else
in a house I've never seen   you would have taken it and made it palatable:

diminutive   hardy   green   I know that things go on
no matter what their form   the red-faced boy I bore   the sweaty fern
outside your door   the volcanic thing aching in the lonely ocean of the night   age-old
atoms stuttering through some kinetic memory event

trying this time for closure   for a formal ascension to place   I have never wanted anything
but the last three symphonies   as loud as possible   searing through the beloved's flesh

YEAR OF THE BEES[1]

January

s	m	t	w	t	f	s
1	2	3	4	5	6	7
8	9	10	11	12	13	14
15	16	17	18	19	20	21
22	23	24	25	26	27	28
29	30	31				

February

s	m	t	w	t	f	s
1	2	3	4	5	6	7
8	9	10	11	12	13	14
15	16	17	18	19	20	21
22	23	24	25	26	27	28
29						

March

s	m	t	w	t	f	s
				1	2	3
4	5	6	7	8	9	10
11	12	13	14	15	16	17
18	19	20	21	22	23	24
25	26	27	28	29	30	31

April

s	m	t	w	t	f	s
1	2	3	4	5	6	7
8	9	10	11	12	13	14
15	16	17	18	19	20	21
22	23	24	25	26	27	28
29	30					

May

s	m	t	w	t	f	s
		1	2	3		5
6	7	8	9	10	11	12
13	14	15	16	17	18	19
20	21	22	23	24	25	26
27	28	29	30	31		

July[1]

s	m	t	w	t	f	s
		1	3	5	7	9
11	13	15	17	19	21	23
25	27	29	31	33	35	37
39	41	43	45	47	49	51

June

s	m	t	w	t	f	s
1	2	3	4	5	6	7
8	9	10	11	12	13	14
15	16	17	18	19	20	21
22	23	24	25	26	27	28
29	30	31				

August

s	m	t	w	t	f	s
					9[2]	

September

s	m	t	w	t	f	s
1	2	3	4	5	6	7
8	9	10	11	12	13	14
15	16	17	18	19	20	21
22	23	24	25	26	27	28
29	30					

October

s	m	t	w	t	f	s
		1	2	3	4	5
6	7	8	9	10	11	12
13	14	15	16	17	18	19
20	21	22	23	24	25	26
27	28	29	30	31		

October[3]

s	m	t	w	t	f	s
		1	2	3	4	5
6	7	8	9	10	11	12
13	14	15	16	17	18	19
20	21	22	23	24	25	26
27	28	29	30	31		

December

s	m	t	w	t	f	s
1		3	4	5	6	7
8	9	10	10[4]	12	13	14
15	16	17	18	19	19	21
22	23	24	25	25	27	28
	29	30	31			

---

1   the year the attic will flood   the year I will learn how to scream   the year my son will be born

2   **A:** July this year will arrive earlier than expected   it had been running on the trails with me   it passed June somewhere by the willows

Q: what was it running from

A: it was running away from a death   a heat death   the death of a language   the death of a friendship   time is folded and so we think if we run fast enough we can get back ahead of such deaths   this is why some people keep photographs   but I have no photographs of July

Q: what will you do in this month   what will you do when it comes around again

A: I was lost and I was cold   I thought you had to do this part of it yourself   the slow walks   the books saying *avoid* and *if* and *spectrum*   the insomniac screams   I was dreamed up by August   reminiscing again   I held burning herbs to my toes and watched the skin strain and jump around like a trapped animal held tight beneath a tarp   I was looking in some unreachable fence at some calm peach body of water

Q: you were hot?

A: I was a spider   I was a daughter   I was sleeping   I was not a single degree closer to myself

3   I sat holding various things over his head   rattles   mobiles   board books   our dog   I wanted him to pay attention to me   I sat taking pictures of the two of us   trying to make it mean something   he was clean   I was making everything so much harder with my brain   I wanted to make it all worth it   (the knives   the screams   the neighbors   the attic)   I was completely static   I wanted to no longer feel like the heaviest body in space   I wanted space   he tore holes in it all the time

4   In the second October you will begin meditation   you will want to remember what it was like to live in the original October   you will meditate on the fields where the fighting happens   you will meditate underground

and in religious revival tents   history books now show how full you were of anger   you sound just like a bird winging over the rivers red with spoiled milk   you were silent in your mind when the fires went off

there were two fires   one in each October   there was a story about an arrow that flew straight into a man smoking something that made him feel disconnected from his own body   the man was aroused   he was thinking of a picture he once saw of a woman who hadn't yet been born

he turned himself into a wildflower and planted himself in the ground   he kept growing   he thought he would just wait until she came around to pick him   hold him against her body   time and the earth

folding over him like giant waves
time and the earth rocking him to sleep

5   **Q**: did you think no one would notice

A: did you notice

Q: was there anything blue then

A: the moon   the new shoes   the
terrible occlusion of vessels   the shining
mess of the flesh that was left   the song
and the song and the song

Q: have you become someone different now   have you forgiven it

A: when I asked about the morning star   it
could not tell me why I shouldn't stand
up and weep   I believe the earth is asleep
in the well-lit past   I grieve it   I relive
everything   I am a whole black greek sky
of reliving   if the world can forgive then
wrap me up in the hot sun of the noon
of December   I will be immune to touch then
I will creep up over the horizon   a waking eye
I will become a mother   I will sleep inside

WE WILL GO HOWEVER FAR IS NECESSARY TO MAKE IT COHERE

The pushing from
the inside makes its own

topography   In your
lifetime you may have to

do this again: take a pre-
existing body   re-shape

it   make it something you
could inhabit   In the future

everyone will be his own
burning bush   aflame but not

consumed   Presumed
outside of the cosmic void

it seems we populate   As a
great great thinker once

said   *Small grammatical
changes in the personal*

*mission statement could create
a new trajectory*   High and

wide and without
precedent   In the future I

will call you *lovely* from
beyond the grave and by

lovely I will mean
a thousand things but

largely guilt   Preparing
to embrace your perhaps

inhuman face   What
world will you grow

used to   octopuses in the
parking structures   white

supremacists in the pictures
with roses around them

icebergs the size of
Delaware   endless endless

thumbs   In your lifetime you may
rise above it: alone   sheathed

in cold   hair made of glass
weaving a wreath out of what

we've detached ourselves
from   throwing it into the

black   orbiting a single small
offering we've left for whatever

we've been tapestried
into   Stepping back to see

it more clearly but seeing
only our own distorted

features   Stepping back
again   And back   And back

METEOR SHOWERS

We have no water and when the water falls it falls in places
that don't need it   the water doesn't remember anything about
being needed   my son asks for *agua* and I think
it is agriculture's problem   so I give it to him   I fill a table full
of water for him   I want to teach him something here that no one else
could know   something about restraint and wanting and the body
never complete or close to soaring   but he can't even tell
someone who or what he is   I fill a table
full of refusal   all the sad people living beneath palm trees
showering for long hours   pouring water from crystal
vases into the street   feeding secret droplets of water to their
strange dogs always barking at the sky as though they could see
something there they didn't like   I keep taking pictures of the sky
so I can remember it this way   our memory of water is of
the whites of dinosaur eyes   swimming towards us in the dark
I have told my son about the dinosaurs   I have sung
him to sleep with a high dry icy voice I heard from the rocks
if you drive the truck underneath the right light you can see
a notch in the hood on the right side where the hail
came down around us in June   I was afraid and I was thinking
of my son but he wasn't anywhere near it   he was hiding
inside my own imagined cave   the rain used to wave at us
like friendly drivers with friendly eyes   sometimes
I think I hear the skies descending but it is phantom rain
a phantom vibration in the pocket or between the skin
of the cheek and the tongue   when the water
used to fall it would fall from familiar places   from bedsides   from
polaroids   from parasites   from noise and the things that make
noise (mouths   boys   entire voids of you)   you
have heard of the Leonids   of the Perseids and their summer
showers hailing from the boy who came out of the wooden chest
in the sea   the radiant comet trailing its debris   when the meteors
used to fall they would fall from familiar places   but these alien
stones   these geminids   come from something defying
classification   not a comet   no halo   no coma   just a hot hot
corpse floating so close to the sun its slow yellow bodies

39

no longer make any sense   my legs are in the wildflowers
my legs are rooting through the dust   my son and I come out
of the dry house   we stand in our swimsuits   dry mouths
opening up   dry clouds parting   we fall down in the streets
we fan the trees and the meteors and the dinosaurs with flame
the mystery is what you have after the body learns its name

IT'S PRETTY FUNNY WHEN YOU THINK ABOUT IT

Whatever you do, don't let the dying
bats get to you. Don't mourn
for vaquita. Don't imagine the dead boys

to be your own; the delicate scaffolding
of daily life couldn't bear it. Have you heard
the one about the world

ending? The one we tell all the time now,
typing it out in tiny boxes, adding
emojis, thinking up variations on

at daybreak? Have you seen the picture
of the golden man's son
sitting in the plaid shirt on the

orchestrated tree stump? God, it's
incredible! I laughed for days. Listen,
I like you. I'd like to invite you to live with me

in a colony simulating life on Mars.
We'd receive water every two months, food
every four. We'd have to wait 20 minutes

to send an email. We'd dream about the
backyards we'd left behind. We'd have to be
cautious, tuning out the echoes

of our old dogs sleeping, filling the silence
with Donna Summer songs. Yes, this
could be the new perfect reality, the one

we escape into before the missiles descend
in the middle of our favorite
Starbucks locations—just the two of us,

41

sleeping in a solar dome on some rocky
surface, growing potatoes out of red
dirt, telling ourselves we'd truly hit on it, the

ultimate escape, the final frontier, the slate
wiped clean for another crack at
utopia, the sun rising so far away—and 39

minutes later than we're used to, 39
minutes in which we could really
make ourselves laugh thinking about

those stupid frogs in those stupid forests,
those creepy octopuses with their
dumb RNA editing, those ridiculous ponds,

green and blue and pink in the light,
humming with those idiotic schools of
silver creatures who never had

the capacity of mind to tear down
their shimmering parts until only
tongues in cheeks remained.

YEAR OF OUMUAMUA[6]

January

s	m	t	w	t	f	s
1	0	1	0	1	0	1
0	1	0	0	1	1	1
0	0	1	1	0	1	0
0	0	1	1	1	1[5]	

February

s	m	t	w	t	f	s
1	2	3	4	5	6	7
8	9	10	11	12	13	14
15	16	17	18	19	20	21
22	23	24	25	26	27	28

s	m	t	w	t	f	s
				1	2	3
4	5	6	7	8	9	10
11	12	13	14	15	16	17
18	19	20	21	22	23	24
25	26	27	28	29	30	31[6]

April

s	m	t	w	t	f	s
1	2	3	4	5	6	7
8	9	10	11	12	13	14
15	16	17	18	19	20	21
22	23	24	25	26	27	28
29	30					

May

s	m	t	w	t	f	s
		1	2	3		5
6	7	8	9	10	11	12
13	14	15	16	17	18	19
20	21	22	23	24	25	26
27	28	29	30	31		

June

s	m	t	w	t	f	s
	1	3	5	7	9	
11	13	15	17	19	21	23
25	27	29	31	33	35	37
39	41	43	45	47	49	51

July

s	m	t	w	t	f	s
1	2	3	4	5	6	7
8	9	10	11	12	13	14
15	16	17	18	19	20	21
22	23	24	25	26	27	28
29	30	31				

August

s	m	t	w	t	f	s
						9

September

s	m	t	w	t	f	s
1	2	3	4	5	6	7
8	9	10	11	12	13	14
15	16	17	18	19	20	21
22	23	24	25	26	27	28
29	30					

October

s	m	t	w	t	f	s
	1	2	3	4	5	
6	7	8	9	10	11	12
13	14	15	16	17	18	19
20	21	22	23	24	25	26
27	28	29	30	31		

s	m	t	w	t	f	s
		1	2	3	4	5
6	7	8	9	10	11	12
13	14	15	16	17	18	19
20	21	22	23	24	25	26
27	28	29	30	31		

November

s	m	t	w	t	f	s
1	1	3	4	5	6	7
8	9	10	10	12	13	14
15	16	17	18	19	19	21
22	23	24	25	25	27	28
29	30	31				

---

5   the ghost horse understands binary code   the first woman who rode him worked on computers for a living she would tap things on his flanks   unconsciously sending a message to herself   *you are alive   you are remembering this*   they rode all over   once she told the ghost horse   *you are more real than any*

horse alive today   he knew it was true   he was a load of blue blue clouds   he was a brute   he had sung his way up out of the earth like a horror movie   when he met the woman he felt as though he had seen her before *memory subjects us to pain*   she tapped   *what are we riding towards if not towards pain*

6   **A:** and you thought you knew   didn't you

Q: how did you find it   (the south pole)

A: first there was a sacrifice   then there was
the ice between us   and the hunger pains
and the nice stars carrying their nice star babies
and finally the compasses pulling downwards
like birds coming in for their lost nests

Q: was there anything blue then

7   the year the ghost horse will rise for the first time   the year we will become paranoid   the year we will magnetize ourselves to the bottom of the globe

THE MYTHS WE INHABIT

It is hard not to feel at least a little abandoned.

I myself am strong enough to run on very
little. A handful of almonds, a few stolen hours
on the couch with the lights off.

The half-slept-in bed becomes a condition of life.

Something unsettling and ecstatic moves
across your field of vision. It is bigger
than you remembered.

Raising your voice, you erase all of your silences.

IV. ENCLOSED

The living body cannot be a self-enclosed world unto itself but must be open to the needs of life. In terms of the species, living beings cannot simply act for self-maintenance and continuity but must in some way become other than themselves in order to have a future.

—Claire Colebrook

I'm a creep, I'm a weirdo. What the hell am I doing here? I don't belong here.

—Radiohead, "Creep"

THE TRUTH IS OUT THERE

I want to have the porcelain skin of the skeptic

lover. I want
to be invited into your hotel room
to see the grainy video tape, episode

after episode. I want to always just miss

the awful thing that happens in the night
in the room with the moon
coming in. I

want a data printout over fax. I want to

be the woman in the room
with a concealed weapon and no visible
cleavage. I also want to wear lipstick.

I want to be a

witness, and then
I want to be the victim in the trunk.
I want you to put the X on your window.

I want to be relieved.

ANOTHER INEXPENSIVE SOLUTION
WITH A BIG PAYOFF

We bought a house recently so now I have a house   I want to stress
that this is not an allegorical house   This is a real live house
One thing you have to do with houses is decorate them   make them
look like they are real places where real people live and love each other and have
"inspiration"   This "inspiration" often takes the form of vases and cleverly arranged
unread books   Did you know that when you are decorating a house
you are also helping to shape its personality   Every house has a personality   Some
are Type As pretending to be "chill"   Some are immigrants   Some
are cool blue pools with nothing inside and no bottom   Some
are anxious to see you leave   Some are sad   withdrawn   high high up
so interested in turning on the fan and closing their eyes   I don't
know how to live in a house or how to inhabit
space   I don't know what shade of tinted primer to use   I am a tall strange
silence with no feet   Everyone keeps telling me about "self-care"   Most
mornings I wake up and I fucking hate myself   which seems positive   it seems
like a step in the right direction   When I was younger I wanted to go
off the grid   to leave society and live in the wilderness in a cabin in
Maine   and now here I am   I just bought a house in the suburbs   I drink
kale smoothies and I like them   I enjoy going to Target   I follow
Kim Kardashian on Instagram ironically   In my house   when you
enter   I want to have a big skull and a defaced portrait of Lee Harvey Oswald
just very casual-like   displayed above a tasteful brown urn full of
baby's breath   I want to fill one room entirely with pieces of confetti
on which I have printed the word "fuck" really tiny   so only a tiny person
could read it   I am probably going to go with neutrals for the living room and
dining room areas   From the backyard I want some spiders to
creep in   They will be wet   They will look like
they have been through a weather event   I want the bedroom crammed
with empty glass cases   I want the kitchen angry and
thick with steam   I know I cannot avoid mirrors   I put on
three layers of skincare products this morning
sensitive skin eye cream   anti-wrinkle moisturizer   tinted sunscreen
I am obsessed with thinking about what The Thing is that will come
into my life and destroy it   will it be cancer   will it be the death of a child
will it be An Accident   It Was Just A Normal Morning and   We Had Coffee

Like We Do Every Morning And Then   will it be the suicide
of a friend   will it be the slow attrition of passion until I can no longer
bring myself to chop avocado   touch my husband   read a novel   laugh
by accident   until I am a slow dark cradle rocking somewhere beneath
the continent   I am trying not to think like this   I have a house to decorate

I WOULD LEAVE MYSELF INSIDE A CLASS CASE IF POSSIBLE

In my fantasy there's an avalanche   we're on top
of the mountain   we believe we're going to die   The press
of ice and yes transgression acting on us   We're inside a tent

sometimes in a crevasse   a cave-like enclosure   There's snow
everywhere and we are cold but we are cold together   In
my fantasy there is some predator in the wood and we have to hold

very still until it passes   our backs to a tree   our breath held
between us like a round quiet ring   You have built this little home
and you have had to be good to it   You built it out of something less

permanent than language   less solid than hands   There was the day
no one could stop watching the news and the first person I thought of
was you   In my fantasy we obsess over the same page of the same book

and we cannot stop talking about it even as outside a lightning storm
approaches   or perhaps a tornado   carrying cows   If you were to build
it again would you exclude me   wall me in   would you be

more careful with the nails   In my fantasy the elevator's stuck
the trunk of the car is locked   you can't get your sunglasses off   the semi
is blocking the entire freeway   we believe in lightning

setting the house on fire   For all your escapist suburban reveries
it's the locking in that absorbs me so completely I could caulk holes in my
walls with it   until not a single pinprick of an out remained

YOU SHOULDN'T WRITE A SONNET
WITH A MACBOOK IN IT

and yet, you try. At first it sounds all wrong;
sonnets are for lips, eyes, vines. (Microchips
can't kiss, tiny cameras can't pine.) The lines
jam up against each other: misfed reams,
machines performing duties skewing off
their programmed script. Tempted to rip it up,
discard it, post it on your blog, slog through
another draft, you cover up the sheet.
Repeat, rewrite. And yet. The light outside
the window shivers. From the page, a sound
of brown and green and twining. Drowned in heat.
If you were to re-read, you'd see a glint
of silver surface quickly swallowed by
a bird. I've heard it hurts to see them fly.

THIS IS AS ACCURATE AS I CAN MAKE IT

There are people now who specialize in drawing elaborate cityscapes
from memory, people who seem to exist only
in time-lapse videos. If I sat down to draw a picture of your face,

it would resemble you exactly, a network of
graphite lines and ovals and curves, a birds-eye view
of some other planet's central hub, dotted with hotels for its

inhabitants, skyways for its cold days. Churned-up soil before
the tulips grow. In my memory you are overly populated,
you smell like piano, you are a map of the transient spaces

that comprise our fulgurating beings. When you look in the
mirror, do you ever have trouble
zooming in to reveal the tiny bodies going about their daily

routines, hailing the cars, steeping the tea, picking the scabs
they've been told not to pick? And all of them in t-shirts and jeans,
humming to mute the buzzing in their heads. Somewhere in the vicinity

of your jawline, a congregation of people ventures
downwards. You only know they are
deep sea diving and not skydiving by the

flippers. When I sit down to draw a picture
of my face, I become aware of the clouds flickering past the window
much faster than they should, hours trespassing

in seconds. The figures sheathed in black descend around you
through the blue. Our bodies: frozen
panoramas; our eyes: intricate temples built to withstand any

sun. Our time-lapse video would look a little
like this: the ice being brought out for the wound, the puddle
in the wings, the hot yellow horizon rushing up like a flame.

DAILY COMMUTATIVE PROPERTY

My son in silhouette against the door.
Sunrise or something else.
The day picks up, the drop-offs happen, we

pass Ernie's Al Fresco every
morning on the 2 and
look out on the mountains just to see

if they've maybe shifted slightly
overnight. A Post-It Note
reminds me: "poem called 'Don't Text Me.'"

Taken in, the faces switch. The
raptor scream of love
unravels in the ear. A singing mass

of steel wings underneath
the slung-out concrete like a
dove. Undone, we sew our chests up,

clasp the buttons,
smooth the sheets. The shoes we wear
make echoes down the streets.

✦

Sunrise or something else. The day
picks up its things & drops
them off. Don't text me; I'm driving, or:

I'm trying not to think
about the words I haven't said. I've
read about a lot of men

pretending to have some sort of control
over the weather. I've combed my
hair out, wet, over the bed. I've looked

out on the mountains. I've been
fed. I've set
all the correct alarms.

✦

The mountains shifted slightly
overnight. My mouth in silhouette,
watching them, waiting. The

heart in particles
weaving through space. The heart
running in place. Something

sings by us. There was
one day I'll never try to stop
remembering. Something other

than sun streaming down my face
and bouncing off the mirror. Don't
text me until later tonight. Tread

carefully, the raptor screams,
the doors might lock behind you.
Some other son might find you.

✦

We pass Ernie's Al Fresco, on his way
to work. Every day, my son:
strapped in and taken

elsewhere. Every day the alarms go
off. I know the men in ties
are doing this, too. I know it's not

a unique burden. My son

touches his belly button, touches
my face. Rapt, I scream. I

pour the cream into the
cup. Text me,
I won't answer. Call

me up, I'll let it ring. The mountains
sing. If I had had a daughter,
I would tell her: Take your heart out.

Turn it into steel.
Now stay that way: chest open,
silver mass a choir in your hands.

ATOMIC LICENSE

We may perform monstrous
rules   We may hold on
to necessary flight

> *We have an interval and then*
> *our place knows us*
> *no more*

When Walter Pater parted
from his lover   certainly
our atoms charted
the space between them

The strange and
lovely problem of
bodies falling from these
numberless windows

> *Our physical life*
> *is a perpetual motion*
> *of them*

Certainly   his lover

passing into our cells
at night

> *And if we continue to dwell*
> *impressions*
> *unstable*
> *flickering*
> *which burn and are extinguished*

We see the ice disks
and assume them
landing targets for
alien ships

it contracts still further:
the whole scope of observation
the narrow chamber of the
   individual mind

We forget already
the interdimensional life
of the crystal beating
in our restless cages

   To burn always with this
   hard   gem-like flame

I don't think you could engineer
a machine to move it
as smoothly

   to maintain
   this ecstasy

like some type of arctic
buzzsaw—

READ THIS WHOLE THREAD

The pictures of monkeys with iPhones are really the definition of everything wrong with our society today. 1/
It's disgusting, we've corrupted everything around us, all the monkeys want to do now is take selfies, they won't even eat the fruit the 2/
zookeepers have been setting out faithfully every morning. And this isn't just the adults. We're talking young monkeys, baby monkeys, 3/
monkeys we suspect to have some monkey form of mood disorder, monkeys who have demonstrated a desire to eat meat, monkeys without any 4/
eyes, monkeys born in the wild, monkeys educated in reading Braille, monkeys I have called friends. 5/
I only bring this up because if you look at what Congress is doing, the ripple effects are clear. I know. I used to be a senate aide. 6/
If you want some practical advice, I can give it to you. Stay tuned. I rarely brag like this, but after all I am the astrophysicist 7/
who wrote the book you're quoting incorrectly from. *No one directs our fate.* I know if you Google that you'll get Hawking, 8/
but honestly, I came up with that on my own. I bet Hawking would hate the Russian plot to revive the wooly mammoths in some 9/
sort of twisted IRL sequel to Jurassic Park. This is the kind of bipartisan cause we can all get behind. Leave the soil and the wooly 10/
mammoths to rest in peace. If I sound passionate about this, I'm sorry. It's just that in my experience (see earlier thread), I saw first 11/
hand the devastation people cause with their pesticide use, their colonies of dead bees, their well-meaning inspections 12/
of remote caves. I'm sorry, I know I'm rambling. If you've made it this far, thank you, and remember not to succumb to the 13/
reigning narrative of apathy. At present, some monkey somewhere may be capturing your image in the background of his selfie, 14/
filtering it, cropping it with care, turning it very gently into some strange new form of narcissistic primate art. 15/
Even now, I have trouble feeling seen. 16/16

I DON'T HAVE AN IPAD

which could be a problem on airplanes now   as they're
slowly getting rid of those screens you've stared at on the backs of seats
while contemplating certain disaster   the crash you've secretly known
would end it your whole life   But why stare in front of you
when you can crane your neck down   lovingly   at your digital
lap   your external pixelated body   We should not worry
about these forthcoming analog airplanes   Nothing
beats the tenacity of the mind   roving over the available
options   convinced it will eventually land on something
that will resolve the unruly flesh around it   like a man
being interviewed by Werner Herzog   unable to
stop the way the camera sees him   talking into the silence
after it should have been over   Is this too
abstract   I have hit on a number of things that I think will make it
better   a ten dollar planner from Target with a cloth cover   a daily
meditation habit   an article on the exercise routines
of famous men   a son   first one
now another   still making fists inside   an idea I had and made
manifest   The rollout is expected to happen gradually   Maybe
this will be a good thing for the women who live in cabins
tucked away in the few remaining spaces of the country without
cell phone towers   the women who suffer from headaches   nausea
itching   vomiting   an inability to decipher numbers written down on a
page   Maybe they will be able to fly now to silent places they've only
ever dreamed of   Have you heard   in any case   about the man who
had the revelation with his hand in the water   I think the articles
are right   too much light in a darkened cabin
can make you restless

EMERGENCY QUEEN

Sometimes, honestly, I'm exhausted
from all the rallying, accosted on all sides
by this eternal series of events
going wrong and then
hanging there, the theories about
whether or not it was worth it, the
earth crumbling around the
edges of the pot in the wrong
pattern. The gray hairs, the
unforeseen exhaustion. I know
why the queens in those hives
gorge themselves on royal jelly,
quelling any minor cell's desire
to give up, to succumb to it. Her life
a lengthening buzz, as much
as seventy times longer than the
worker bees around her. Freeze a day,
multiply it, try to guess how someone else
could ever know the length of your
particular hour. If a queen
dies unexpectedly
the bees in this queenless colony will build
new large and somewhat
slapdash cells, to produce an "emergency
queen," who's usually smaller, less
prolific. It's true,
all the leading immortalists had fathers
who died young. All started out in tech. All
are men. All love the sound of *never*: never
sag, never ache, never drag, never
break. The waking always the same: the
clock check, the supplement regimen, the
underseasoned food. Who'd have thought
living forever could be so little fun? The
hyperbaric chamber out of the way,
behind the couch. The mouse in the lab,
crouched over, not looking as spry

as you'd hoped. The slope of senescent
cells, the hum of the rough knee.
To be free from the minutes. *Life is but
a shadow: the shadow of a bird on the
wing.* The bird above, singing. The
ringing in your ears. You walk the maze
you made, you get lost in it. The time
of your life is now. Now the gift
and the punishment. Now the slow
sand spooling out its minute knives
on your hand. Now the land, stretching
out like the pool that birthed you, blue
and warm and boxed in, so enclosed. *Mox
nox.* I want to add
wishes. Look at my shadow and you
will see your life. The youthful profile
wavers in the wind, smells like dried
flowers, *spigelia marilandica,* a
hummingbird favorite. All this nightly
torpor and so little rest. I would build you a
nest that never unraveled. *Tempus edax
rerum.* Lie here with me on the porch
and listen to the dogs bark
until the queenless beings around us
begin to cement the cells that will
usher in the new era, that will begin the
beautiful and possible descent.

V. ABLAZE

There is one term of the problem which you are not taking into account: precisely, the world. The real. You say: the real, the world as it is. But it is not, it becomes! It moves, it changes! It doesn't wait for us to change ... It is more mobile than you can imagine. You are getting closer to this reality when you say as it 'presents itself'; that means that it is not there, existing as an object. The world, the real is not an object. It is a process.

—John Cage

Keep a little fire burning; however small, however hidden.

—Cormac McCarthy, *The Road*

MATINEE

It was a windy day and I wasn't sure when I might next get an afternoon to myself, so I decided to go see a movie. I bought a bottle of water, I brought my poncho in case I got cold. One often gets cold at the movie theater. There were a lot of previews, and the guy behind me said *So many previews* to his wife in the kind of annoyed voice that tells people you are annoyed at something obviously happening right in front of everybody's faces. He was very easy to hear as it was only me, this couple behind me, and another couple way in the back, and both of the couples were easily octogenarians. In fact, I hadn't seen another person my age in the whole place. Just me and the octogenarians. In a horror movie, maybe there would have been some plot revolving around this—all the elderly people in town kidnapping all the younger people in the town out of jealousy or spite or just for kicks, because where do you get your kicks as an octogenarian if you're not kidnapping people in their 20s and 30s, maybe holding them in moldy basements for a while just as a lark, feeding them old bits of ham from your refrigerator. Then going to the movies, which would turn out to be the source of new victims. I imagined how such a film would end—in a movie theater, of course, the final girl running down the aisles in her nightgown, setting the screens on fire as she ran, bursting through the flames for a triumphant confrontation with the evil manifested in these bodies all around her. It would have to be a spectacular finish. The catch, of course, would be that she, too, was aging, and you'd see her in the sequel, grinning in the dark, a single candle next to her, having turned into the thing she'd thought she'd defeated. I took a sip of my water. The movie had already started. On the screen, a small boy who I swear looked just like my son was eating a hamburger. I think I started to cry right away. All these tiny relics of the body, and where do they go? Something that used to be skin, papery, disintegrating in the soft recesses of our carpets. I could feel my face refusing to match the shape of the face I'd thought I had. *So many previews*, I could hear the man behind me still saying, *So many previews*. But the movie has already started, I said, the matches in my purse growing cold beneath the seat.

A RICH HISTORY OF MEN PAINTING THINGS BLUE

Things rise up all the time even if you can't see them   once a member of the
Winnebago tribe owned this land   he lived in your house with his
body   he touched his wife here   they fought   they thought they could create
something lasting   they had a horse they would ride up the hill
& at still points in the afternoon they would talk about light   about the absence
of sound   they made films together when the films were flat things sucked
clean of pigmentation & speech or song   they sometimes longed
to burn it all down   smoke drifting up from Edendale   like tiny blurry
birds all floating together   like a dead gray ghost whooshing up & out of the body
things rise up all the time   the last native speaker of the man's tribal language
died in 1856   twenty years before he was born   a whole generation of people
walking around with a heavy silence where a tongue should be   today you think
you will plant something   digging in the ground you find a horse hitch   rusted
full of a history you weren't a part of   *it was probably his horse*   you think   &
you remember the fight you had with your wife   last night   she was stone faced
she said she *understood*   but she didn't understand anything   the way these films
rise up inside you all the time   ending in fire or earthquake or the death of your son
& you yourself standing there with your hands full of wildflowers watching the smoke
rise up around the city   if you lie down in the dirt right now & look straight up
you will see nothing at all   you recall a rich history of men painting things blue
there are whole planets devoted to men like this   from behind   an invisible horse
nudges you   breathes on you with its hot horse breath   you want nothing more
than to ride it down the hill silently like a tiny blurry bird   both of your tongues
making sounds out of some desert film   some tribal dance   you rise up all the time
even though nobody sees you   you rise up all the time

THAT WHICH IS FORM IS EMPTINESS,
THAT WHICH IS EMPTINESS FORM

It was a summer of extractions, evacuations.

No eyes, no ears, no nose, no tongue, no body, no mind.

When the woman woke up in the dark room she was hollow.

The colorless sky outside the window.

They do not appear or disappear, are not tainted or pure.

Once, she had couches to lie on for entire evenings. Once, she believed in a correlation between purity and ecstasy, invisibility and some ultimate transcendence—

all shall be well and all shall be well and all manner of thing shall be well.

The woman was hungry.

The woman knew the exact moment a part of her turned into you, became something monstrously whole.

You should have been a part of me forever.

I don't know why, in the Bible, it's Adam who loses a rib.

It has always been the woman who gives up a piece of her body in order to create the other,

the woman who lives with a hole in herself for the rest of her life while the thing she created waltzes around talking to snakes.

And you, my son, purifier, occupier, already awake in another room without me, rephrasing yourself in chaotic syntax.

You, vomiting up the hourglass with an image of my former face at the bottom, mouth open, waiting for the erasure, grain by grain.

You, rebuilding your house of my flesh somewhere not yet on fire.

THESE WEREN'T A FEW BIRDS

By the time you see the man lighting the cigar   it's already too
late   The invisible particles already floating up towards
ignition   The cuts to the face   and the face   and the face   A couple
frames shorter every time   You refresh the browser window
hoping this headline will be the one to crystallize your silent and
impossible fantasies
                      Unnatural geometry   Ordinary animal   By the time
the stream lights up the horror will have already passed   The horror that's
a feeling   The horror that's a premonition   The horror that happens
when the kids are in the bath and you receive the final text   The horror
that's the face in the mirror   The horror that's the unknown   The
horror that's the realization you didn't make the
                                    call   The cuts
to the face   and the face   and the   The horror that's the call coming from
after the attempt   The horror that's the body   The body that's on
fire   The fire that's a smile with nothing behind the eyes   The invisible
particles floating towards whatever happens after the unutterable return
into this life   The man lighting the memory   The memory of the sound
of bees   It's already too late
                      A couple frames shorter every

time

FROM A DISTANCE IT COULD BE
A SLOWLY BLOOMING FLOWER

Zoom in: burning birds
set beautifully in
motion   Spiraling
into poetry's trap

The city stayed in its
tent and wrote poems
for 1600 years   The city

was consciously
artistic   operating outside
the plot

Zoom in again: nothing
resembles light
except a sonnet on fire
(mistaken for a bird)

THE RESEARCH WAS
RECENTLY PUBLISHED

The world today would seem to call for a poetry of
    facts, a poetry of
testimony. The news feeds do a good enough job
    displacing you into a surreal
landscape, a planet built of language making its own
    reality, *weather extremes* replacing *climate*
change as though a shift in the way we
    say something
will also shift our experience of that thing, and,
    surprise, it will, you will look
up one day and the sky will be filled with nothing you
    were ever responsible for, nothing you
could ever have been asked to consider,
    and you will find yourself craving
a burger and fries
    in the middle of the street
covered with the small shedded hairs of babies
    who can't question anything
at all. Here is a fact: in the square of sun
    cast by the sliding glass door
at the back of the house
    you can feel a certain substance
slipping away from you. Too easy to say
    time, too precious to say
teleology. Another fact: the colors seem to have been
    manipulated, heightened and faded
at once. As for testimony, I have little to give you.
    I can tell you the elbows
weren't as soft as they looked. That
    the only reason I didn't duck my head
was anger. That if you could see my mouth, you'd see it
    open, empty, filling up with
fire. Have you read about the worms
    who inherited
from scientifically-starved parents

a genetic readiness for
starvation, a DNA bracing for hunger? Another fact:
    our brains can keep
re-molding themselves in someone else's vision
    far past the point of
logic, which is something to be afraid of
    and to celebrate. The
beautiful consuming machine is setting
    because we said so.

IN THE POST-WAR ERA, IMAGES OF WHEAT FIELDS WERE COMFORTING

It's an old adage: what doesn't kill you
warms you up inside. The fires we try
to escape, the fires we light in anger,
the fires we harbor quietly inside white
rooms. There are fires from stars
that died billions of years ago
still flaring up across the void into
our telescopes. If we left now,
some vain mission to stop the
pain, we'd arrive only to find the
embers long gone, the former star
a tiny neutron entity. In this reality,
I suppose, we are astronauts, or at
least people with access to a
vehicle capable of exiting the Earth's
orbit. We are probably cold; you can't
light a fire on a rocket ship. I am not sure
how many of us there are. Five, or ten.
Maybe there are a handful of rockets,
all on the same futile trajectory,
launched out of the moral obligation
to stop needless death wherever
possible (even star death) or out of the
selfish desire to get just a little closer
to something that burns so hot and so
bright. As we are rising up in the night,
still so close to the surface of the Earth,
we pass over a golden wheat field
undulating in the dark. It looks like no one
has ever walked through it. No light source
in sight, a shadow falls across it, this
place unchained from time and
composed out of color only, no form.
The figure weaving out slowly from the
silent farm house pushes her way

into the field, the rough stalks
whispering up against her cheeks,
drawn towards the shadow hovering
in the middle of it all as though cast
from the bottom of a trap door to the
stars. As she lies down beneath it, we
lose sight of her. Did she wait
for whatever drifted above to pause
and lift her up? Did anybody hear her?
Did she stay there, counting the minutes,
until the stark American morning
dawned at last, and she discovered herself
turned overnight into a body
of resignation? We move closer to the
fire and it moves farther away from
us. We wonder what it feels like
to shroud yourself in golden stalks
emerging from the earth. We don't
know what to say about the dew
collecting everywhere but here.

NOTES

IT'S WEIRD AND PISSED OFF, WHATEVER IT IS takes its title from a quote by the character Clark in John Carpenter's *The Thing* (1982).

THE PEOPLE WHO LIVE IN BOATS was partly inspired by the story of the Whanganui River, which the Maori tribe fought to have awarded legal status as a living entity. It takes a line from Gerrard Albert, lead negotiator for the Whanganui iwi [tribe].

"DID YOU HEAR THAT, THIS GUY THINKS IT'S BEEN TERRIBLE" was said by Donald Trump, mockingly, to an aide, reinforcing his view that the early stages of his presidency had gone quite well.

ATOMIC LICENSE quotes from the Conclusion to Walter Pater's *The Renaissance: Studies in Art and Poetry* (1873).

EMERGENCY QUEEN uses some language from traditional mottos from sundials in both Latin and English.

THAT WHICH IS FORM IS EMPTINESS, THAT WHICH IS EMPTINESS FORM adopts some language from the Heart Sutra, a sutra in Mahāyāna Buddhism. It also quotes the 14th century anchorite Julian of Norwich.

THESE WEREN'T A FEW BIRDS is indebted in some of its imagery and title to Hitchcock's *The Birds* (1963).

IN THE POST-WAR ERA was written after seeing the John Rogers Cox painting *Wheat Field* (c. 1943), which I thought, from a distance, was a 1940s version of a UFO painting, and couldn't shake that image from my head.

78

ACKNOWLEDGMENTS

Grateful thanks are due to the editors of the following publications, where these poems appeared in print for the first time:

The Cincinnati Review: "Meteor Showers," "Weak Stars"
Coda Quarterly: "Daily Commutative Property," "The Truth is Out There"
DIAGRAM: "You Shouldn't Write A Sonnet With A MacBook In It"
Hobart: "It's Weird and Pissed Off, Whatever It Is," "It Works in Any Space, From Bedrooms to Kitchens," "How Come No One on Twitter is Talking About This," "It's This Again," "Another Inexpensive Solution With a Big Payoff"
The Georgia Review: "Eventually an Expert Came, Hooded Her, and Took Her Away"
The Missouri Review: "Something Not Empty"
The Offending Adam: "From a Distance it Could be a Slowly Blooming Flower," "Atomic License"
Painted Bride Quarterly: "Housewarming," "The People Who Live in Boats," "Emergency Queen"
The Timberline Review: "Situations of Casual Danger"

And, of course, thanks to the many many people who have made this manuscript possible by reading, talking, editing, or simply being there. To Genevieve Kaplan for being generous about reading series invitations. To Andrea Quaid and the Bard Institute of Writing and Thinking for shimmering. To the Aldous Huxley cabin on the Goji Berry Farm in Taos, New Mexico, for being a space where space could happen. To Carrie and Jay Wilcox & Christa and Patrick Cantwell for the support and the extra childcare and the love. To Jess and Suzi and Michelle and Greg for the friendship and for letting me be me. To my students for the daily inspiration and challenge and perspective and acceptance and laughter. To Valerie Wallace for choosing this manuscript, and Cati Porter & the Inlandia Institute for making it real. To Leo for needing to be alive. To Cooper for being my ultimate joy and forever archenemy and the perfect future shelter-seeker. And finally to Chris—for standing with me in the sun and building whatever it is we're living inside of right now, for keeping the emergency backpacks up to date, and for believing in whatever's on the other side of now.

Elizabeth Cantwell is a poet and high school Humanities teacher living in Claremont, CA, with her husband and their two sons. Her poetry has appeared in a variety of journals and publications. Her first book of poems, *Nights I Let the Tiger Get You* (Black Lawrence Press, 2014), was a finalist for the 2012 Hudson Prize; she is also the author of a chapbook, *Premonitions* (Grey Book Press, 2014).

ABOUT THE HILLARY GRAVENDYK PRIZE

The Hillary Gravendyk Prize is an open poetry book competition published by Inlandia Institute for all writers regardless of the number of previously published poetry collections.

HILLARY GRAVENDYK (1979-2014) was a beloved poet living and teaching in Southern California's "Inland Empire" region. She wrote the acclaimed poetry book, *HARM* from Omnidawn Publishing (2012) and the poetry collection *The Naturalist* (Achiote Press, 2008). A native of Washington State, she was an admired Assistant Professor of English at Pomona College in Claremont, CA. Her poetry has appeared widely in journals such as *American Letters & Commentary, The Bellingham Review, The Colorado Review, The Eleventh Muse, Fourteen Hills, MARY, 1913: A Journal of Forms, Octopus Magazine, Tarpaulin Sky* and *Sugar House Review*. She was awarded a 2015 Pushcart Prize for her poem "Your Ghost," which appeared in the Pushcart Prize Anthology. She leaves behind many devoted colleagues, friends, family and beautiful poems. Hillary Gravendyk passed away on May 10, 2014 after a long illness. This contest has been established in her memory.

ABOUT INLANDIA INSTITUTE

Inlandia Institute is a regional non-profit and literary center. We seek to bring focus to the richness of the literary enterprise that has existed in this region for ages. The mission of the Inlandia Institute is to recognize, support, and expand literary activity in all of its forms in Inland Southern California by publishing books and sponsoring programs that deepen people's awareness, understanding, and appreciation of this unique, complex and creatively vibrant region.

The Institute publishes books, presents free public literary and cultural programming, provides in-school and after school enrichment programs for children and youth, holds free creative writing workshops for teens and adults, and boot camp intensives. In addition, every two years, the Inlandia Institute appoints a distinguished jury panel from outside of the region to name an Inlandia Literary Laureate who serves as an ambassador for the Inlandia Institute, promoting literature, creative literacy, and community. Laureates to date include Susan Straight (2010-2012), Gayle Brandeis (2012-2014), Juan Delgado (2014-2016), Nikia Chaney (2016-2018), and Rachelle Cruz (2018-2020).

To learn more about the Inlandia Institute, please visit our website at www.InlandiaInstitute.org.

OTHER HILLARY GRAVENDYK PRIZE BOOKS

Former Possessions of the Spanish Empire by Michelle Peñaloza
Winner of the 2018 National Hillary Gravendyk Prize

Our Bruises Kept Singing Purple by Malcolm Friend
Winner of the 2017 National Hillary Gravendyk Prize

Traces of a Fifth Column by Marco Maisto
Winner of the 2016 National Hillary Gravendyk Prize

God's Will for Monsters by Rachelle Cruz
Winner of the 2016 Regional Hillary Gravendyk Prize
Winner of a 2018 American Book Award

Map of an Onion by Kenji C. Liu
Winner of the 2015 National Hillary Gravendyk Prize

All Things Lose Thousands of Times by Angela Peñaredondo
Winner of the 2015 Regional Hillary Gravendyk Prize

www.ingramcontent.com/pod-product-compliance
Lightning Source LLC
Chambersburg PA
CBHW081140090426
42736CB00018B/3417